INTRODUCTION TO FOOD SCIENCE

A Kitchen-Based Course

AN OVERVIEW

SECOND EDITION

TEACHER'S PLANNER AND ANSWER KEY

INTRODUCTION TO FOOD SCIENCE

A Kitchen-Based Course

AN OVERVIEW

SECOND EDITION

Dale W. Cox

TEACHER'S PLANNER AND ANSWER KEY

AN **EDIBLE KNOWLEDGE**® SERIES COURSE

Copyright © 2024 Dale W. Cox. All Rights Reserved

All rights reserved. No part of this publication may be reproduced, distributed, or transmitted in any form or by any means, including photocopying, recording, or other electronic or mechanical methods, without the prior written permission of the publisher, except in the case of brief quotations embodied in critical reviews and certain other noncommercial uses permited by copyright law. For permission requests, write to the publisher, addressed "Attention: Permissions Coordinator," at the address below.

ISBN: 978-1-948515-31-3

Library of Congress Cataloging-in-Publication Data is available.

Published by Beakers & Bricks, LLC

Cover design by Glen M. Edelstein and Dale Cox

Photographs by Dale W. Cox, unless otherwise specified in Pictures and Illustrations Attributions
Edible Knowledge® logo art by LeAnne Cox
Edible Knowledge® is a registered trademark of Beakers & Bricks, LLC
Printed in the United States of America

Beakers & Bricks, LLC
PO Box 1014
Asheboro, North Carolina 27204
www.edible-knowledge.com

TABLE OF CONTENTS

PROPOSED SCHEDULE — *1*

CHAPTER 1
FOOD SCIENCE BASICS — *3*
VIDEO LINKS 3
JOURNALING IDEA 4
CHAPTER REVIEW 4

CHAPTER 2
A SCIENCE PRIMER — *5*
WHAT DO YOU THINK? 5
JOURNALING IDEA 5
CHAPTER REVIEW 6

CHAPTER 3
FOOD PROCESSING — *7*
VIDEO LINKS 7
WHAT DO YOU THINK? 8
JOURNALING IDEA 9
CHAPTER REVIEW 10

CHAPTER 4
PROXIMATE ANALYSIS — *11*
WHAT DO YOU THINK? 12

JOURNALING IDEA 13
CHAPTER REVIEW 13

CHAPTER 5
WATER — *15*
VIDEO LINKS 15
WHAT DO YOU THINK? 16
JOURNALING IDEA 16
CHAPTER REVIEW 16

CHAPTER 6
CARBOHYDRATES — *18*
VIDEO LINKS 18
WHAT DO YOU THINK? 19
JOURNALING IDEA 19
CHAPTER REVIEW 19

CHAPTER 7
PROTEIN — *21*
VIDEO LINKS 21
WHAT DO YOU THINK? 22
JOURNALING IDEA 22
CHAPTER REVIEW 23

CHAPTER 8
FAT (LIPIDS) — *25*
JOURNALING IDEA 25
VIDEO LINKS 25
CHAPTER REVIEW 26

CHAPTER TESTS
JUST THE QUESTIONS   29

- CHAPTER 1 TEST - FOOD SCIENCE BASICS 30
- CHAPTER 2 TEST - A SCIENCE PRIMER 32
- CHAPTER 3 TEST - FOOD PROCESSING 34
- CHAPTER 4 TEST - PROXIMATE ANALYSIS 36
- CHAPTER 5 TEST - WATER 38
- CHAPTER 6 TEST - CARBOHYDRATES 40
- CHAPTER 7 TEST - PROTEIN 42
- CHAPTER 8 TEST - FAT 43

CHAPTER TESTS
THE ANSWERS   45

- CHAPTER 1 TEST - FOOD SCIENCE BASICS 46
- CHAPTER 2 TEST - A SCIENCE PRIMER 49
- CHAPTER 3 TEST - FOOD PROCESSING 52
- CHAPTER 4 TEST - PROXIMATE ANALYSIS 55
- CHAPTER 5 TEST - WATER 58
- CHAPTER 6 TEST - CARBOHYDRATES 61
- CHAPTER 7 TEST - PROTEIN 64
- CHAPTER 8 TEST - FAT 65

PROPOSED SCHEDULE

2 INTRODUCTION TO FOOD SCIENCE: AN OVERVIEW, TEACHER'S PLANNER & ANSWER KEY

16 Week Schedule	Week 1	Week 2	Week 3	Week 4	Week 5	Week 6	Week 7	Week 8	Week 9	Week 10	Week 11	Week 12	Week 13	Week 14	Week 15	Week 16
Chapter 1 - Food Science Basics	■															
Chapter 1 Company Worksheets		■														
Chapter 2 - A Science Primer			■													
Chapter 2 *What Do You Think?*, *Journaling Idea*, and *Chapter Review*				■												
Chapter 3 - Food Processing					■											
Chapter 3 *What Do You Think?*, *Journaling Idea*, and *Chapter Review*					■											
Chapter 4 - Proximate Analysis					■											
Chapter 4 Proximate Analysis Worksheets					■											
Chapter 4 *What Do You Think?*, *Journaling Idea*, and *Chapter Review*						■										
Chapter 5 - Water						■										
Experiment 5-1 (2 weeks)							■	■								
Experiment 5-2 (2 days)								■								
Chapter 5 *What Do You Think?*, *Journaling Idea*, and *Chapter Review*									■							
Chapter 6 - Carbohydrates									■							
Experiment 6-1 (1 day)										■						
Experiment 6-2 (5 days)											■					
Chapter 6 *What Do You Think?*, *Journaling Idea*, and *Chapter Review*												■				
Chapter 7 - Protein												■				
Experiment 7-1 (2 days)													■			
Experiment 7-2 (1 day)													■			
Experiment 7-3 (1 day)														■		
Chapter 7 *What Do You Think?*, *Journaling Idea*, and *Chapter Review*														■		
Chapter 8 - Fat															■	
Experiment 8-1 Mayonnaise Separation (1 day)															■	
Experiement 8-2 Shortening Change of State (1 day)															■	
Chapter 8 *What Do You Think?*, *Journaling Idea*, and *Chapter Review*																■
Bonus Section																■
Bonus Section *What Do You Think?* and *Journaling Idea*																■

CHAPTER 1
FOOD SCIENCE BASICS

VIDEO LINKS

Page 12

Batter and breading process:
https://www.youtube.com/watch?v=TXx0Ig8RY6w

Page 13

General food extrusion:
https://www.youtube.com/watch?v=2544pagWns

Cooker extruder:
https://www.youtube.com/watch?v=TiaCrn23g1k

Pasta extruder:
https://www.youtube.com/watch?v=TXtm_eNaIwQ

4 INTRODUCTION TO FOOD SCIENCE: AN OVERVIEW, TEACHER'S PLANNER & ANSWER KEY

JOURNALING IDEA

Imagine that you're a professional food scientist having just completed 20 years of working for several food companies. Write a two-page fictional account of your favorite new product development project. Include details such as who you worked with, what types of challenges you had when developing the food, what the package looked like, and how successful it was in the marketplace. Use complete sentences, and share your story with someone else.

> Grading Hints: The student learned about the product development process in this chapter, including the team that works together to produce new products. This includes marketing, research and development, engineering, and production, among others. Formula development in the lab is followed by scale up in a production facility, followed by development of specifications that ensure the quality of the finished product.

CHAPTER REVIEW

1. Look through the cupboards in your pantry or kitchen, as well as the refrigerator and freezer. Inspect the packages to see who's selling the product. Evaluate at least five different products and make a list, including the company names that produce them. You can find the company information on the side panel, and often on the container front.

> Grading Hints: self-explanatory

2. Research two food companies and their websites online (Nestlé, Kellogg's, or General Mills, for example). You can choose any that interest you. If you can't think of any, go to your kitchen cupboard and look at the side panel of something in there. A section will indicate "distributed by" and give a company name. Use the forms on the following pages to record information about the companies, the products they make, and job opportunities that may exist there.

> Grading Hints: self-explanatory

CHAPTER 2
A SCIENCE PRIMER

WHAT DO YOU THINK?

Who is the most famous scientist you can remember? What was that discovery or research that made that person famous? Do you think there are many more things to discover?

> Grading Hints: The answers to these questions will be as varied as the individual answering them, with the exception of the last one: there is a whole world to discover yet, and many of the things we believe are true by current scientific understanding will be further refined and sometimes even disproven by new research and more ingenious scientists...maybe that's you!

JOURNALING IDEA

Why do atoms react together the way that they do? Describe—in as much detail as you can—the difference between "how" and "why" in scientific knowledge and progress.

Sometimes people, even scientists, use the phrase *settled science* to refer to a body of knowledge that's believed by them to be proven and complete. Do you agree with this? Why or why not? Use complete sentences.

> Grading Hints: The answers to this questions will be as varied as the individual answering them, with the exception of the last one: there is a whole world to discover yet, and many of the things we believe are true by current scientific understanding will be further refined and sometimes even disproven by new research and more scientific testing!

CHAPTER REVIEW

Write short essay responses to each of the following situations:

Imagine you're working in a laboratory where standard and metric measurements are both used. Describe what precautions you'd need to take to avoid making a mistake.

Search on the internet and read about the NASA loss of a Mars orbiter that occurred in September 1999 (http://www.cnn.com/TECH/space/9909/30/mars.metric.02/). Describe what it must have been like for those on the project to realize the mistake after years of work and tens of millions of dollars spent to bring the project to fruition.

Mars Lander Link

> Grading Hints: This is a real problem in science. In the food industry it can and has happened. That's why in processes with critical, hard to measure ingredients, prepackaged, already portioned portions are used wherever possible.

CHAPTER 3
FOOD PROCESSING

VIDEO LINKS

Page 40

    Wheat harvesting:
    https://www.youtube.com/watch?v=zX8K1OpBCj4

    Smaller-scale wheat cleaner:
    https://www.youtube.com/watch?v=FFfZ3dzUXHs

Page 43

    Sugar from sugar beets:
    https://www.youtube.com/watch?v=VRZX1bAnbes

    Sugar from sugar cane:
    https://www.youtube.com/watch?v=jCKt02NGjfM

Page 44

Gelatin manufacturing:
https://www.youtube.com/watch?v=uf0uEWGWLgg

Page 47

Hammer mill:
https://www.youtube.com/watch?v=e6trUtoIOZE

https://www.youtube.com/watch?v=Y93rf2cNloI

Page 48

Making shredded wheat:
https://www.youtube.com/watch?v=oWFFXuhvkKA

WHAT DO YOU THINK?

1. Give five examples of a minimally processed food that you eat at home.

   > Grading Hints: Examples of minimally processed foods are carrots, broccoli, and other raw vegetables.

2. What are the five most highly processed foods currently in your house?

   > Grading Hints: Examples are frozen dinners and sweet tries like Hostess Twinkies® or Ding Dongs®. Of course, there are many examples that fall between minimally and highly processed foods.

3. Why do you think that the term processed food has such a negative connotation?

   > Grading Hints: This is subjective, but in the author's opinion, some of the negative connotation is earned, some not. The best thing about learning about it is you can know what to be worried about and what is OK.

4. Now that you know more about processed food, do you think it deserves the negative view that it has? Why or why not?

> Grading Hints: A subjective question again.

5. How can you educate yourself about the processed food debate's pros and cons? Talk to someone about what you think and see if they agree.

> Grading Hints: This course is a great place to start your learning! The author's hope is that every student will realize they need to do their own research, from multiple sources, and use their own brain to determine their stance. Then, don't be afraid to change your mind when new data becomes available. This goes for everything, not just the processed food debate.

JOURNALING IDEA

Now that you have a better understanding of what minimally and highly processed foods are, pretend that you're currently 70 years old and you've lived your life eating mostly highly processed foods. Describe what you look like, and how you feel.

Next, pretend that you're 70 years old again, but this time you've lived your life eating mostly minimally processed foods or foods made at home, only consuming highly processed foods when you needed the convenience or wanted a treat. Describe what you look like, and how you feel.

Which person do you want to become? Write down some goals about how you want to eat and how you'd like to feel as you grow older. Discuss these goals with someone close to you.

> Grading Hints: The general idea here is that you will be healthier eating as close to nature as possible, using more highly processed foods when you need the convenience or want a treat.

CHAPTER REVIEW

Write short essay responses to each of the following questions:
1. Name a food that you have processed before it was ready to consume and describe this process. Example: you may have washed, peeled, then cut a carrot into bite size pieces.

> Grading Hints: The example should get them thinking. Other examples are apples (wash, cut, remove the cores and seeds), or meat of various types, that needs to be cooked to kill any pathogens.

2. Describe in detail the most and least processed food you consumed over the last 5 days.

> Grading Hints: Remember, even the raw carrots are processed to some degree (washed, maybe peeled, etc.)

CHAPTER 4
PROXIMATE ANALYSIS

Correction: Pages 54 and 55 of the 2nd Edition show a Calculated Proximate Analysis chart. An Eagle Eyed student found an error which will be corrected in subsequent editions. The Protein line incorrectly converts 5.0 grams protein to 1.0 gram, when no conversion is needed. Below is the corrected chart, with a *huge thank you* to all our customers who take the time to let us know of errors they find:

Calculated Proximate Analysis

	Units Used On Nutrition Facts Side Panel (g or mg)	From Nutrition Facts	Convert to grams (if needed)	Calculated Percentage
Total Fat	g	1.0	1.0	1.7%
Total Carbohydrate	g	46.0	46.0	78.0%
Ash (add mineral values together)	mg	600.0	0.6	1.0%
Protein	g	5.0	5.0	8.5%
Water: (serving size) minus (fat+carbohydrate+ash+protein)			6.4	10.8%
			Total Percentage	100.0%

WHAT DO YOU THINK?

Look at the Nutrition Facts panel for five of your favorite foods and complete the Calculated Proximate Analysis tables on the following pages. Try to choose foods that are different from each other—for example, a breakfast cereal, mixed frozen vegetables, a candy bar, a carbonated beverage, and hot dogs. It's likely that some foods you choose may not have all the items that are listed on the worksheet. There may also be foods that have more than what's listed. Manufacturers have some flexibility on the label, and sometimes want to accentuate that a food does contain a certain nutritional component. Do the best you can.

After you've completed filling out the forms, answer the questions below.

1. How do the foods differ from one another?

> Grading Hints: For example, bread will have a lot of carbohydrate, but not very much carbohydrate as sugar, which is a carbohydrate itself. A breakfast cereal will have also have a lot of carbohydrate, but a great percentage of it will be as sugar.

2. Think about what you eat in a typical day. Are you mostly getting a lot of one component, such as carbohydrate or fat? The diet of a typical U.S. citizen is high in refined carbohydrates such as sugar or finely milled flours.

> Grading Hints: Answers will depend on the student's diet.

3. Pretend that today is your first day on the job as a food scientist and you've been asked to design a food which is similar to one of those you've just analyzed. What might be your next step?

> Grading Hints: Look at the ingredient line, find out what the proximate analysis makeup is of those ingredients, generate your "best guess" formula based on what you know, then calculate what the side panel would look like. If it is in the ballpark, you likely have a good place to start. The next is to begin making prototypes based on your formula and make adjustments as needed.

JOURNALING IDEA

Ask an adult you know if their doctor recommends they avoid certain foods, or components of a food, such as sodium. Write about how using the Nutrition Facts side panel can help with this task. What about foods prepared at home? Note: Some questions in this workbook, including ones below, require thinking beyond the material presented in the workbook. Be creative and use your mind...you can do it!

> Grading Hints: The side panel nutrition information tells you what is in the food, including substances that perhaps you have been told to avoid by your doctor. For food prepared at home there are websites and books that can help you determine what a recipe's nutrition information will be. Some recipe sources now list this information as well.

CHAPTER REVIEW

Write short essay responses to each of the following questions:

1. How can the Nutrition Facts found on labels help you plan your diet? Do a couple of searches on the internet and write about what information used to be on labels before the Nutrition Labeling and Education Act of 1990.

> Grading Hints: Knowing approximately the amount of macro food components you are eating, including carbohydrates, protein, fat, and further if the carbohydrates are sugars or starch, combined with the recommended intake of these things, can help you understand if adjustment is needed. If you have health concerns, the side panel can also help you avoid things that are currently understood to be harmful for certain conditions. An example is sodium intake and those afflicted with high blood pressure.

2. What would you do if packaged food didn't contain nutrition information?

> Grading Hints: Some smaller packages are not labeled for individual resale and you may not have the original box in which the package was housed. Other countries also

> have different or non-existant labeling laws. In either case, a call to the manufacturer will answer the question for you.

3. Write about the school experience of an adult you know. Did they go to college? If so, did they graduate in the major they chose when they began? Are they still working in that same field? Write about what this may mean for your own life.

> Grading Hints: I am hoping by now the student has understood that a degree is a starting point, not a slot in which you must work the rest of your life. A diversion from the original plan can start right away, or even decades down the road.

CHAPTER 5
WATER

VIDEO LINKS

Page 69

Ice formation and hydrogen bonds:
https://www.youtube.com/watch?v=UukRgqzk-KE

Exploding frozen carbonated beverage cans:
https://www.youtube.com/watch?v=RGhgpJAHsT8

https://www.youtube.com/watch?v=WFyaL6iozKY

WHAT DO YOU THINK?

We've only scratched the surface of all the interesting things we know about water and the role it plays in processing food. The Edible Knowledge® workbook Introduction to Food Science: Water discusses this topic in much greater detail. Can you think of any food that doesn't use water, or have water in it to least some degree?

> Grading Hints: I can't think of even one example. All that I know use water in production and to some degree contain water even though it might be to a small extent.

JOURNALING IDEA

Imagine what life would be like without water. Write a short science fiction story about a new colony on Mars and how the colonists had to recycle their water. Include methods and invent names for them, and specifically explain how you'd clean the water that's reclaimed from human waste, showers, and washing activities. Write about how you might be able to capture water that escapes through evaporation when preparing food.

> Grading Hints: Have fun with this assignment. Water is so important to all life, thinking about how to preserve it, even in cooking, can be fun. For example, using a lid, or a hood that will condense water vapor.

CHAPTER REVIEW

1. A building footing, or the solid foundation on which a building is constructed, must be deeper below ground in cold climates than in warmer climates. Why?

> Grading Hints: The ground will freeze if the weather is cold enough, but only to a certain depth. When water freezes in the ground, ice expands, causing the ground to shift. Footings must be below the "frost line" where the ground will freeze, otherwise buildings will shift and crack. Hydrogen bonding is amazingly strong.

2. Your new food company, ABC Foods, would like to sell frozen ice cream that has organoleptically undetectable ice crystals. What might be your biggest hurdle, and why?

> Grading Hints: There could be a couple valid answers here. Keeping ice crystals small as they are forming requires skill and equipment. For a finished ice cream, keeping them small in a freezer over time is a bigger challenge. Ingredients that can be used to "stabilize" the ice cream, preventing ice crystals from getting too large, are important.

3. Think of the items around you that are solid, including the plastic in the pen you may be holding, the graphite in the pencil, or the plastic in the keyboard you may be using. Are these components frozen? Explain.

> Grading Hints: They are. Anything in a solid state is "frozen", although usually only water is spoken of as being ice in this state. The solid can be different states, which is beyond this course.

4. When a large wave approaches the shoreline of an ocean, the water immediately in front of it will pull back, or recede, into the ocean. This phenomena is noted distinctly when a tsunami (tidal wave) occurs. From what you've learned, you should be able to explain one of the primary causes.

> Grading Hints: Hydrogen bonds are what holds water together.

CHAPTER 6
CARBOHYDRATES

VIDEO LINKS

Page 84

Maillard Browning:
https://www.youtube.com/watch?v=c7WI41huAok

Page 85

Non-Enzymatic (Maillard) Browning:
https://www.youtube.com/watch?v=c7WI41huAok

Page 88

Starch granules swelling as they hydrate and gelatinize:
https://www.youtube.com/watch?v=L6vYxYE1jOg

WHAT DO YOU THINK?

Look at the side panel of packaged foods in your kitchen. Do you see any that contain modified starch? This will usually be cornstarch. What role do you think the modified cornstarch plays in the food? Call the toll-free number listed on the package and ask the manufacturer what type of modified starch is in the food, and what role it plays. You may or may not get an answer right away. However, you'll eventually get some type of an answer if you indicate that you really would like to know! For example, if the person doesn't know, ask them to contact you when they find someone who does know. Sometimes companies won't share exactly what's in their formula because it's the secret to their success, and they may think you're a competitor trying to copy them!

> Grading Hints: Modified food starch can play many roles, usually to help preserve desired textural characteristics longer than an unmodified starch would allow.

JOURNALING IDEA

Now that you know a little about carbohydrates, write about your favorite one carbohydrate molecule. Why is it your favorite? Describe what you can about its structure and what you know about that structure that may determine the characteristics you enjoy. Write about something you can make in the kitchen that may have a similar structure, and therefore that you may also like.

> Grading Hints: Most peoples favorite "carbohydrate molecule" is a sugar of some kind. Sucrose is the most common as it is table sugar and is important in desserts.

CHAPTER REVIEW

Write short essay responses:
1. Why does a dry cornstarch film look and feel so much like plastic wrap?

> Grading Hints: Cornstarch is made up of amylose and amylopectin, large polymers. Plastic is also made of polymers. Edible films are used for some types of packaging and are made of plant polymers.

2. Describe a starch granule's thoughts as if it were alive while being boiled, gelatinized, and fully hydrated, and ultimately becoming part of butterscotch pudding. This was the starch granule's life goal.

> Grading Hints: This was a bit of fun, although I should probably have put the "goal" part first. The first part is a little scary!

3. Write a short story about your favorite mostly carbohydrate food, including why it's your favorite. How much fat or protein does it contain that help it to be your favorite carbohydrate?

> Grading Hints: Choose a food and use your proximate analysis skills to pick it apart.

4. Describe one area of your life that doesn't involve chemistry in some way.

> Grading Hints: This is a bit of a trick question. There is no area of your life that doesn't involve chemistry in some way. Your physical body is 100% chemicals, and chemical processes are what keep you alive.

CHAPTER 7
PROTEIN

VIDEO LINKS

Page 105

An interesting video discussing enzymes and denaturation:
https://www.youtube.com/watch?v=Z_ZieKjEQ7s

Page 115

Enzymatic Browning:
https://www.youtube.com/watch?v=Tt9FYHmM0jU

WHAT DO YOU THINK?

What about the color changes in the meat? It was red and became gray or brown depending on the preparation method. What's going on there? Have you ever reheated cooked meat? It doesn't taste anywhere close to freshly prepared meat. What about meats that can be stored in a bag at the checkout line? Why can they sit like that, without refrigeration? Why are bologna and some hot dogs pink?

What other enzymes are used in food production? What role do they play? Do we have to eat animal protein in order to survive?

As you can tell, we've just scratched the surface of proteins in food science. This chapter was only intended to give you a taste (pun intended) and perhaps stimulate some curiosity and a desire to find out more!

> Grading Hints: These questions are somewhat rhetorical, meaning I don't really expect that the student can answer them all right now. They are designed to get you thinking. There is much to learn about proteins and how they interact to become a good food scientist.

JOURNALING IDEA

How many different ways can an egg prepared? Ask a cook you know or do a web search regarding the question and write down what you find. How is it possible that the same egg, when prepared differently, can have textures such is found in boiled egg whites or scrambled eggs, or even a hard or soft foam such as meringue on a pie? How can your new knowledge of protein help explain this?

> Grading Hints: Eggs are prepared in innumerable ways. Depending on the experience of the cook consulted it can be a long or short list. The student should be able to understand and explain how different processing techniques affect even the same starting protein ingredients, including eggs and especially egg whites. Examples: Whipping into a foam and then baking or just cracking an egg into a hot pan into a hot pan result in quite different finished foods. Whipping denatures the white through mechanical action prior to baking, whereas frying denatures the white and sets the structure through heat, without mechanical action.

CHAPTER REVIEW

Write short essay responses:

1. What is your favorite protein source and why? How do you like that protein prepared? Describe what you can regarding the science behind what happens to the protein as it's prepared and finally becomes the food you enjoy.

> Grading Hints: The favorite protein source will be student dependent. Example: dry roasted beef. As the beef cooks the collagen fibers (a protein) denature and if cooked long enough become gelatin like. The result is fully cooked muscle fibers that are fall-apart tender since collagen is what holds them together.

2. American citizens eat a lot of beef, which comes from cattle. Cattle eat only plants. How do cattle convert plants into protein? Can humans do this?

> Grading Hints: Cattle and other ruminant animals can live solely on limited plants, such as grass, through fermentation that happens in an additional stomach. Humans can't do this, but with careful selection are able to get most needed proteins needed through plants and plant products as well.

3. An effective way to sterilize something is to expose it to a radioactive source, which will effectively kill any living process. Spices are often treated this way to reduce microbial activity, since heating them would destroy, or at least change, their flavor. From your understanding of proteins, can you think of why radiation is so effective? Hint: Radiation of the proper type denatures proteins.

> Grading Hints: I give this one away in the hint. Proteins are integral to life, all the way down to the cellular level. If cellular proteins cease to function and can't be replaced, the organism, large or small, will shortly die. This is a primary reason why radiation "poisoning" can be lethal if it is of the right type and duration.

4. Describe what you think the scientific mechanism may be for the different results seen in preparing beef from Protein Experiment #3. Can you think of other proteins where a similar mechanism might be employed to achieve a desired result?

> Grading Hints: Protein denaturation plays a role in almost every food using proteins. Egg whites, whole or ground muscle foods, and milk proteins when making cheeses or yogurts are all examples.

5. Think of an animal muscle. Tendons, muscle fibers, and other connective tissues are all mostly protein, but they look and act quite differently. How can this be when they are all made of protein?

> Grading Hints: The amino acid sequences are different, but these are all examples of proteins.

CHAPTER 8
FAT (LIPIDS)

VIDEO LINKS

Page 133

Making mayonnaise by hand.
https://www.youtube.com/watch?v=moz_zNPdbhI

JOURNALING IDEA

Write about your favorite food. Break it down according to its proximate analysis. Is it a healthy food, or something that should be consumed in moderation? Write about why it's your favorite food. Include descriptions of how it makes you and your mouth feel when the food is consumed. Think and write about how, as a food scientist, you'd try to remove all the fat or sugar from the food and still make it your favorite food. It's hard!

> Grading Hints: Proximate analysis refers to breaking the food down into it's component parts: carbohydrates, proteins, fats, water, and ash. The student should be able to break down their favorite food into percentages of these five components. Ideas on how to change the amounts of ingredients will depend on their creativity, and at this point I wouldn't expect too much, but want to get them thinking.

CHAPTER REVIEW

Write short essays in response to five of the following questions:

1. How might science continue to evolve regarding any of the subjects you learned about in this book?

> Grading Hints: I make the point in the text that science is never settled but continues to change with new experimentation and understanding.

2. How do your language classes, such as studying English or Spanish, help prepare you for a career in food science? Is this different than if you decided to pursue a career in medicine?

> Grading Hints: The text references experiences I had working in different countries. Language skills can help significantly. The same is true for other professions, although in different ways.

3. Doughnuts are fried in fat that's solid at room temperature. Why might this be important? Describe a doughnut fried in a fat that's liquid at room temperature.

> Grading Hints: A donut fried in room temperature liquid fat would be oily and have a mouthfeel most would find unpleasant. If you were to squeeze the donut, oil would be visible to some degree, like squeezing a sponge. Room temperature solid fat won't do this.

4. Chocolate (mostly fat) that's considered premium melts at a temperature just below what's normal in your mouth. Less expensive chocolates will incorporate fats or other ingredients that raise the chocolate melting point. Describe what effect this may have in the mouth.

> Grading Hints: The mouthfeel of cheap chocolates with a higher melting point feel waxy in the mouth. A way to mimic this is toe suck on a piece of ice, then put a piece of normal chocolate in your mouth. The mouthfeel is unpleasant.

5. You may have heard that a particular food is a *good source* of protein, or some other component that's beneficial for the body. What does this mean? What's the definition of a "good source?" How might your new food company, ABC Foods, develop a food product that health conscious people might want to consume?

> Grading Hints: "Good source" is a United States FDA term meaning the food contains at least 10% of the recommended amount of the nutrient. New foods can be developed for customers looking for a certain nutrient.

6. Continuing from the last question, think of three products ABC Foods can develop and describe some challenges associated with their development and formulation. How long would the shelf life be? What would the name be? What would the packaging look like?

> Grading Hints: For example, I worked on Cream of Wheat, fortifying it with calcium and iron to the "excellent source" level. Gluten free foods. Good source of calcium macaroni and cheese. Fatty ingredients might reduce the shelf life due to oxidative rancidity. High sugar content foods can absorb moisture readily. Packaging and can help slow down moisture transmission and rancidity reactions. These are just examples.

7. What have you found most interesting about your introduction to food science? What have you found least interesting?

> Grading Hints: Completely up to the student.

8. Did all the experiments you conducted work as expected? If not, why didn't they?

> Grading Hints: Determining why an experiment doesn't work can be some of the best education. I do reference the most common reasons for failure in the discussion portions.

9. Evaluate the ingredients of real butter. Examine each ingredient. Does butter contain sugar? Protein?

> Grading Hints: Butter has naturally occurring sugar (lactose) and protein present in cream.

10. How might your mathematics courses help you as a food scientist?

> Grading Hints: Creating sometimes complex spreadsheets were an almost daily activity during my profession. Statistics calculations, while mostly accomplished through software, needed to be understood as well.

> Grading Hints: Bonus Section
> The answers to these questions depend on the student and their chosen university's website.

CHAPTER TESTS
JUST THE QUESTIONS

The following pages contain multiple choice tests associated with each chapter in the course. If you have access to the online version of this course, these tests are included and automatically graded. You may wish to make copies of these pages for the student to take as each chapter is completed. The next chapter includes the questions, answers, and some explanation.

CHAPTER 1 TEST - FOOD SCIENCE BASICS

1. When speaking of a food product, what's a store brand?
    A. A type of product that stores like to buy.
    B. The brand of store you like to shop in.
    C. Also called private label brands, these products are manufactured for a company that will then sell it under their branding.

2. What does avoiding "burning bridges" mean?
    A. Leaving a company on friendly terms.
    B. The interaction between R&D and Quality Control personnel at a food manufacturing facility.
    C. A process conducted by food engineers when designing a new piece of equipment.

3. What is one thing a food scientist specializing in the sensory sciences does?
    A. They use their senses more than most.
    B. They combine statistics and food science knowledge to test food prototypes to give meaningful data to food product developers.
    C. They eat food and use their senses.

4. As a newly graduated food scientist you are employed at ACME Food Company as a Quality Control Specialist. Which of the tasks below might fall within your work responsibilities?
    A. Working with product development scientists to write specifications for a new product.
    B. Design testing procedures to ensure food products are meet the desired quality specifications.
    C. Tasting the food products made at ACME regularly to make sure they meet quality standards.
    D. All of the above.

5. Which United States of America government agencies are involved in enforcing laws associated with the food we eat? Choose all that apply.
    A. The Food and Drug Administration (FDA)
    B. The United States Department of Agriculture (USDA)
    C. The Department of Food, Meat, and Safety (FMS)

6. A general explanation of an extruder would be:
   A. A machine that can make pasta and breakfast cereal.
   B. A machine that flattens and then folds dough.
   C. A machine that forces material through a die (a small, shaped opening).

7. Speaking scientifically, when a food product is evaluated it means that:
   A. It is examined in every way that is important to its final quality. This can include appearance, texture, taste, and aroma.
   B. Determine the cost so that it can be sold profitably.
   C. Make sure that the product tastes the way it should.

8. If you are mechanically minded and like to solve real world, everyday problems, you might consider a career in:
   A. Food science
   B. Food engineering
   C. Quality assurance

9. If you are quiet and detail oriented and like to be by yourself, you might enjoy a job as a food scientist specializing in research & development.
   True
   False

10. As a food science professional I can (choose all that apply):
    A. Work for a well-known food company.
    B. Form my own company and develop my own food products.
    C. Work with people who are talented tasters.
    D. Analyze food products.
    E. Have an easy life without having to continually update my education.

CHAPTER 2 TEST - A SCIENCE PRIMER

1. Choose the best description of mass vs. weight.
   A. Mass is big, but weight can be small.
   B. Mass doesn't change, but weight changes depending on gravity and other forces.
   C. Mass interacts with weight to make objects move.

2. Are hydrogen bonds weaker than covalent bonds?
   A. Yes.
   B. No.

3. Is science "settled", or in other words, not open for new evidence to revise what is currently understood?
   A. Yes.
   B. No.

4. Choose the answer that is most acidic.
   A. Tomato juice, pH about 6.5.
   B. Bleach, pH about 13.
   C. Orange juice, pH about 3.3.

5. What system of measurement is most common around the world?
   A. Roman
   B. English
   C. Metric
   D. Anglican

6. What is a molar solution?
   A. Molar = moles of solute in 1 kilogram of solvent
   B. Molar concentration = moles of solute divided by solution volume

7. When something is described as "basic", in scientific terms it means:
   A. It is simple.
   B. It is acidic and has a high pH.
   C. It has a high pH.

8. Why must scientists be concerned about the metric vs Standard systems of measurement?
    A. A number is a number, but when attached to something important, like speed or distance, a mix-up can prove disastrous.
    B. They really don't need to be concerned, they just need to be careful and choose one or the other.
    C. It depends on where you live whether this should be of concern.

9. Bonding that takes place between molecules of water is an example of:
    A. covalent bonding
    B. ionic bonding
    C. hydrogen bonding

10. What might best describe a water molecule:
    A. Polar
    B. bi-polar
    C. non-polar

CHAPTER 3 TEST - FOOD PROCESSING

1. Are there any foods not processed in some way?
   A. No
   B. Yes

2. A fast way to separate particulates from a liquid that would normally settle over time is called _____.
   A. centripetallation.
   B. centrifugation.

3. From what is gelatin manufactured?
   A. gelatin desserts.
   B. sugar cane stalks.
   C. connective tissues from animals.

4. What does functionality refer to in food science?
   A. What a food component does in a particular situation.
   B. The mathematical description, or function, of an ingredient.

5. What is the author's food philosophy?
   A. Dale eats as close to nature as possible as often as he can, but enjoys more highly processed foods on occasion as a treat or when convenience is needed.
   B. Dale never eats processed food because he was scared off of them after working in the food industry.

6. Select the common sources of sucrose (normal table sugar):
   A. sugar beets
   B. sugar cane
   C. honey
   D. corn

7. A method designed to get the sugar out of sugar beets by floating thin slices in water with movement in the opposite direction of the water is called:
   A. cross-current flow
   B. collateral-current flow
   C. counter-current flow

8. What are common sources of gelatin?
   A. pork skins
   B. animal connective tissue
   C. plant material

9. The following are steps in shredding wheat (select all that apply):
   A. cooking at or near boiling in water
   B. tempering
   C. smashing
   D. shredding by forcing between a smooth and a grooved roller

10. A successful breakfast cereal manufacturer must:
    A. be careful to avoid overcooking the material
    B. make large volumes of cereal in a short time, otherwise they won't make enough money.

CHAPTER 4 TEST - PROXIMATE ANALYSIS

1. What are the main components of all foods?
   A. Starch, roughage, protein, fat, and ash.
   B. Water, fat, fiber, carbohydrate, and ash.
   C. Water, carbohydrates, protein, fat, and ash.

2. How do you calculate how much water is in a food product from the side panel Nutrition Facts?
   A. Add up everything else and subtract it from the serving size.
   B. Weigh it before and after microwaving for 10 minutes.
   C. Subtract the total of the carbohydrates and protein from the total weight.

3. Nutrition Facts available on the side panel can help me:
   A. Avoid eating too much fat.
   B. Figure out if the food is going to taste good.

4. Ash is what is left once all organic material has been incinerated.
   True
   False

5. How do you calculate the amount of carbohydrates in a food?
   A. Add up everything else and subtract it from the total weight.
   B. Feed it to the author and see how much weight he gains.
   C. Determine the amount of sugar (carbohydrate) in the food.

6. Proximate analysis means to:
   A. Water is of primary importance.
   B. use the side panel to analyze a food completely.
   C. "approximately" analyze all food components to find out what's in them
   D. break a food down into it's five basic components.

7. A common element in protein used in laboratory tests to quantify how much protein is present in a food is:
   A. Carbon
   B. Nitrogen
   C. Oxygen

8. Fat is soluble in:
   A. hot water
   B. salty water
   C. some alcohols
   D. ice cream

9. How many milligrams are in a gram?
   A. 10
   B. 100
   C. 1000

10. Using the side panel on packaged foods can help me understand what's in my food and know if I am eating too much of some component that is not good for me.
    True
    False

CHAPTER 5 TEST - WATER

1. As oil solidifies slowly, it's density _____.
   A. stays the same.
   B. decreases.
   C. increases.

2. As water slowly solidifies (freezes), the density _____.
   A. stays the same.
   B. decreases.
   C. increases.

3. Water activity is the partial pressure of water over a sample divided by the partial pressure of water over pure water. Choose the best option below you could use if you were trying to describe this principle to someone.
   A. Some food is really wet, and some is dry. You have to be careful when mixing them together.
   B. Some foods hold onto their water more tightly than other foods. If two products are put together that don't have the same water activity, eventually water will move from one to another, sometimes with a poor result.
   C. All things being equal, high water activity foods hold onto their water, and low water activity let go.

4. What property of water makes it so good at cleaning?
   A. The water activity.
   B. It's bipolar nature.
   C. The equilibrium achieved when things are dirty.

5. Choose the best description of freezer burn.
   A. Concentration of sugars, flavors, and oils, and loss of water at a food's surface, resulting in changes in texture.
   B. When you put your tongue on a frozen popsicle, it gets stuck, and you pull it off, burning your tongue.
   C. Freezer burn makes food organoleptically yummy.

6. Flash freezing describes a very rapid freezing process. Compared to a slow freezing process, ice crystal size from flash freezing would be _____.
   A. larger.
   B. smaller.

7. The Dew Point is the temperature at which air is saturated with water or, in other words, the air can hold no more water.
   A. True
   B. False

8. When making cooked food you intend to freeze, what are some things you need to consider. Mark all that apply.
   A. Concentration of non-water components and unexpected flavor changes that might result (unexpected chemistry).
   B. Changes in texture.
   C. Disruption of raw plant cells.
   D. Freeze-thaw cycles common in today's freezers if it is going to be stored for a long time.

9. All other variables, such as pressure, as water changes phase, such as from a solid to a liquid, or a liquid to a gas, the temperature _____ until the phase change is complete.
   A. increases
   B. decreases
   C. stays the same

10. When an emulsion, such as mayonnaise, is frozen, usually the emulsion will _____.
    A. break, meaning it will separate into its oil and water phases.
    B. become homogenous, blending its oil and water phases.
    C. remain the same.

CHAPTER 6 TEST - CARBOHYDRATES

1. All sugars have the same sweetness.
   A. True
   B. False

2. Choose the best answer that describes fructose, which is more hygroscopic than sucrose.
   A. Fructose picks up moisture out of the atmosphere more readily than sucrose.
   B. Sucrose picks up moisture out of the atmosphere more readily than fructose.
   C. You shouldn't use fructose in most food formulations, because it is hygroscopic.

3. What's the primary difference between sugars and starches? Choose all that apply.
   A. Sugars are sweeter than starches.
   B. Sugars are longer chains of glucose than starches.
   C. Starches are longer chains of glucose than sugars.
   D. Sugars are carbohydrates and starches are not.

4. When bread crust browns and gives us those unique flavors we associate with baked bread, it is an example of:
   A. A good bread cook.
   B. Maillard browning.
   C. A call for butter and honey.

5. Maillard browning requires a _____ and a _____.
   A. protein, reducing sugar
   B. protein, non-reducing sugar
   C. starch, sugar

6. Amylose is a _____ molecule than amylopectin.
   A. Smaller
   B. Larger

7. Hydrocolloids are large molecules that absorb moisture, keeping a food moister, mimicking fat, and providing body and texture.
   A. True
   B. False

8. Examples of hydrocolloids are:
   A. xanthan gum, gum arabic, and cornstarch
   B. xanthan gum, gum arabic, and guar gum
   C. inulin, xanthan gum, and amylose

9. The primary difference between caramelization and Maillard Browning is:
   A. How the reactions start since many of the end products (brown colors and flavors) are the same.
   B. One requires protein, and the other requires sugar.
   C. They are not different.

10. Cooked, hydrated potato starch gels are described as "long". This means that relative to most cornstarches there is _____ amylopectin.
    A. Less
    B. the same amount of
    C. more

CHAPTER 7 TEST - PROTEIN

1. Protein levels of structure include:
   A. Simple, complex, and quaternary
   B. primary, secondary, tertiary, and quaternary.
   C. nitrogen, enzymes, and substrate.

2. The loss of structure in a protein due to heat, for example, is known as _____.
   A. desaturation.
   B. denaturation.
   C. consternation.

3. Glutenin + gliadin, with water and mechanical action results in a new protein complex called _____.
   A. gluten.
   B. glue.
   C. bread.

4. Apples browning and onions producing chemicals that make our eyes water is an example of _____.
   A. nature in action.
   B. enzymatic action.

5. Are gluten-free foods better for everyone?
   A. Yes, because gluten causes problems for human intestinal tracts. That's why there are so many gluten free foods now.
   B. No. Only a select few particular diseases or sensitivities need to be concerned about foods that contain gluten.

CHAPTER 8 TEST - FAT

1. If you wanted a high fat food to be less oily at room temperature, you would want the fat to be _____.
   A. liquid.
   B. fully hydrogenated.
   C. full of cis fats.

2. Another way to source a fat that is more solid at room temperature is to choose one that has _____ fatty acid chains.
   A. longer.
   B. shorter.

3. All fats are built on a _____ molecule.
   A. Large
   B. Chained
   C. Glycerol

4. As fats age they can develop off-flavors and odors that most find offensive. When this happens the fat or food (containing fat) has become _____.
   A. rancid.
   B. extruded.
   C. nasty.

5. The following are ways that can reduce the onset of oxidative rancidity (choose all that apply):
   A. Store in a cool place.
   B. Protect from light.
   C. Flush with oxygen gas to help slow down the process.

6. Generally, donuts should be fried in a fat that is solid at room temperature.
   A. True
   B. False

7. The author believes we should avoid all trans fats.
   A. True
   B. False

8. How many fatty acids does a triglyceride have attached to the glycerol molecule?
   A. One
   B. Two
   C. Three
   D. Four

9. Potato Chips should be fried in fat that is solid at room temperature.
   A. True
   B. False

10. The mucus membrane irritating compound that results from fat thermal decomposition is called:
    A. acrimony
    B. acrolein
    C. smoke

CHAPTER TESTS
THE ANSWERS

The following pages contain the multiple-choice tests from the previous chapter, this time with the answers indicated in ***bold italics***. A short explanation follows each quest. This explanation is intended to further help the student understand the course material.

CHAPTER 1 TEST - FOOD SCIENCE BASICS

1. When speaking of a food product, what's a store brand?
    A. A type of product that stores like to buy.
    B. The brand of store you like to shop in.
    C. *Also called private label brands, these products are manufactured for a company that will then sell it under their branding.*

Store brands are an important part of food manufacturing. Many product brands don't have any manufacturing capability but hire other companies to use their excess manufacturing capacity. This arrangement helps both companies, although sometimes a manufacturing company might be making products that directly compete with their own brands!

2. What does avoiding "burning bridges" mean?
    A. *Leaving a company on friendly terms.*
    B. The interaction between R&D and Quality Control personnel at a food manufacturing facility.
    C. A process conducted by food engineers when designing a new piece of equipment.

To your boss on leaving a job: " I've found a better job, and boy am I glad, because this place was killing me! I couldn't stand it, or YOU, for even one more day! Goodbye!".

"Burning bridges" is never a good idea, and not just because it's not nice or professional. Even if you really don't like a job and can't stand your boss, you should always try your best to leave a job on friendly terms. You were gainfully employed for a while, even if it was a terrible job, and you never know when you might be working with someone again...the world is smaller than you think!

3. What is one thing a food scientist specializing in the sensory sciences does?
    A. They use their senses more than most.
    B. *They combine statistics and food science knowledge to test food prototypes to give meaningful data to food product developers.*
    C. They eat food and use their senses.

Sensory scientists can specialize in lots of areas. Helping product developers figure out which way to go (too salty, or not salty enough, for example) with food prototypes is one of them.

4. As a newly graduated food scientist you are employed at ACME Food Company as a Quality Control Specialist. Which of the tasks below might fall within your work responsibilities?
   A. Working with product development scientists to write specifications for a new product.
   B. Design testing procedures to ensure food products are meet the desired quality specifications.
   C. Tasting the food products made at ACME regularly to make sure they meet quality standards.
   D. *All of the above.*

Food scientists working in the quality assurance area are busy and important professionals in any food production facility. They often design and ensure proper execution of testing techniques to ensure products fall within acceptable ranges and will even help determine what these acceptable ranges may be. Sometimes they may be unpopular as they may make the final decision on whether to throw out production or to let it go to market. They are responsible and respected positions.

5. Which United States of America government agencies are involved in enforcing laws associated with the food we eat? Choose all that apply.
   A. The Food and Drug Administration (FDA)
   B. *The United States Department of Agriculture (USDA)*
   C. The Department of Food, Meat, and Safety (FMS)

The third choice is a fictitious organization I made up, any resemblance to any actual organization is coincidental.

6. A general explanation of an extruder would be:
   A. A machine that can make pasta and breakfast cereal.
   B. A machine that flattens and then folds dough.
   C. *A machine that forces material through a die (a small, shaped opening).*

Extruders are used for all kinds of things, including making pasta and cereal. There are cold extruders and extruders that directly cook and expand products, but they all have in common one thing: they force material in one way or another through a hole which, in extrusion terms, is usually called a die.

7. Speaking scientifically, when a food product is evaluated it means that:
   A. *It is examined in every way that is important to its final quality. This can include appearance, texture, taste, and aroma.*

B. Determine the cost so that it can be sold profitably.
C. Make sure that the product tastes the way it should.

When I was writing this course my use of the term evaluation confused my editor. She was not accustomed to the way I was using it, which confused me as I had hardly ever used the term in any other way. As a result, we added this definition which I hope helps you!

8. If you are mechanically minded and like to solve real world, everyday problems, you might consider a career in:
   A. Food science
   B. <u>Food engineering</u>
   C. Quality assurance

Based on what you read in this chapter, food engineering would be the most correct response. As a food engineer you will use your mechanical expertise to devise and repair machines on a continual, daily basis. You may get to do some of that in the other two professions, but as a food engineer, that is likely your day-to-day job.

9. If you are quiet and detail oriented and like to be by yourself, you might enjoy a job as a food scientist specializing in research & development.
   <u>True</u>
   False

I have known all types of researchers, most of whom, me included, are not the flamboyant life of the party type. However, there are some. In no field of employment do you get to be completely by yourself. There is interaction with other people at least some of the time.

10. As a food science professional I can (choose all that apply):
    A. <u>Work for a well-known food company.</u>
    B. <u>Form my own company and develop my own food products.</u>
    C. <u>Work with people who are talented tasters.</u>
    D. <u>Analyze food products.</u>
    E. Have an easy life without having to continually update my education.

As a food scientist you can pretty much do whatever you want. However, ANY job requires updating your skills or you will find yourself out of a job or outpaced by your competitors.

CHAPTER 2 TEST - A SCIENCE PRIMER

1. Choose the best description of mass vs. weight.
   A. Mass is big, but weight can be small.
   B. <u>*Mass doesn't change, but weight changes depending on gravity and other forces.*</u>
   C. Mass interacts with weight to make objects move.

Mass doesn't change, but weight does depend on where it is, and the forces applied to it. The classic example is an astronaut on the moon vs. the earth. The astronaut has the same mass but weighs less on the moon than on the earth.

2. Are hydrogen bonds weaker than covalent bonds?
   A. <u>*Yes.*</u>
   B. No.

Hydrogen bonds are weaker than covalent bonds, but when many are combined, they can have a great effect. Integral to the behavior of water, hydrogen bonding is important in many food science principles.

3. Is science "settled", or in other words, not open for new evidence to revise what is currently understood?
   A. Yes.
   B. <u>*No.*</u>

Science is NEVER "settled". A good scientist is always open to hearing new evidence that will alter understanding of natural phenomena.

4. Choose the answer that is most acidic.
   A. Tomato juice, pH about 6.5.
   B. Bleach, pH about 13.
   C. <u>*Orange juice, pH about 3.3.*</u>

A lower pH means increased acidity.

5. What system of measurement is most common around the world?
   A. Roman
   B. English
   C. <u>*Metric*</u>
   D. Anglican

The metric system is by far the most common measurement system throughout the world.

6. What is a molar solution?
   A. Molar = moles of solute in 1 kilogram of solvent
   B. *Molar concentration = moles of solute divided by solution volume*

While confusing, this is the way it is and while we won't really use these much in this course, we will touch on molarity and molality from time to time. If you get stuck, come back and take a look at this lesson.

7. When something is described as "basic", in scientific terms it means:
   A. It is simple.
   B. It is acidic and has a high pH.
   C. *It has a high pH.*

"Basic" in chemical terms means a high pH. Examples of basic compounds are baking soda dissolved in water and sodium hydroxide.

8. Why must scientists be concerned about the metric vs Standard systems of measurement?
   A. *A number is a number, but when attached to something important, like speed or distance, a mix-up can prove disastrous.*
   B. They really don't need to be concerned, they just need to be careful and choose one or the other.
   C. It depends on where you live whether this should be of concern.

Especially in today's internationally connected world, great care must be taken to ensure everyone is using the same measurements, or at least converting them correctly. Otherwise, you may end up with an interplanetary experiment crashing and burning.

9. Bonding that takes place between molecules of water is an example of:
   A. covalent bonding
   B. ionic bonding
   C. *hydrogen bonding*

Hydrogen bonding is everywhere in food science. You will learn a lot about hydrogen bonding as it applies to food as you work your way through the course.

10. What might best describe a water molecule:
    A. Polar
    B. *bi-polar*
    C. non-polar

A water molecule is a bi-polar molecule, resulting in it having a negative and a positive side. This characteristic is critically important in our foods as you will see.

CHAPTER 3 TEST - FOOD PROCESSING

1. Are there any foods not processed in some way?
   A. *No*
   B. Yes

Fruit is probably the closest thing to unprocessed. You can pick an apple right off the tree and eat it straight away, but it's good practice to process it a little bit by washing it first. The idea is to not be afraid of processed foods, but to only consume them in moderation when you need convenience.

2. A fast way to separate particulates from a liquid that would normally settle over time is called _____.
   A. *centripetallation.*
   B. centrifugation.

Centrifugation is the same process used for many things, including separating blood platelets from blood plasma.

3. From what is gelatin manufactured?
   A. gelatin desserts.
   B. sugar cane stalks.
   C. *connective tissues from animals.*

While not a pleasant thought, gelatin is derived from connective tissue from animals.

4. What does *functionality* refer to in food science?
   A. *What a food component does in a particular situation.*
   B. The mathematical description, or function, of an ingredient.

Functionality is the term used by food scientists to describe what an ingredient provides or accomplishes in a food. For example, eggs' function in many baked foods is to provide protein structure.

5. What is the author's food philosophy?
   A. *The author eats as close to nature as possible as often as he can, but enjoys more highly processed foods on occasion as a treat or when convenience is needed.*
   B. Dale never eats processed food because he was scared off of them after working in the food industry.

While it might seem presumptuous to state my own OPINION through a test question....well, it is presumptuous. Still, I feel strongly about this as it has made my life easier to not be subject to every new food fad or trend. I do recognize that this is my OPINION, and encourage you to form your own based on what you learn.

6. Select the common sources of sucrose (normal table sugar):
   A. *sugar beets*
   B. *sugar cane*
   C. honey
   D. corn

While honey is a great source of fructose, it is not used to manufacture sucrose, and neither is corn.

7. A method designed to get the sugar out of sugar beets by floating thin slices in water with movement in the opposite direction of the water is called:
   A. cross-current flow
   B. collateral-current flow
   C. *counter-current flow*

Counter-current flow is used in many food manufacturing processes. By exposing fresh beet slices to water with the most dissolved beet juice, and beet slices that have been in water the longest (those where most of the juice has already been transferred into the water) to the water with the least dissolved beet juice.

8. What are common sources of gelatin?
   A. *pork skins*
   B. *animal connective tissue*
   C. plant material

While something approximating the functionality of gelatin can be made from plants, and is even called "plant gelatin", it is not common and is not gelatin. Almost all gelatin used in the world today comes from animal sources.

9. The following are steps in shredding wheat (select all that apply):
   A. *cooking at or near boiling in water*
   B. *tempering*
   C. smashing
   D. *shredding by forcing between a smooth and a grooved roller*

Shredding wheat is one of the simplest ways to prepare wheat for consumption.

10. A successful breakfast cereal manufacturer must:
    A. be careful to avoid overcooking the material
    B. *make large volumes of cereal in a short time, otherwise they won't make enough money.*

Any type of consumer product such as breakfast cereal usually requires making a lot of it in a short time, otherwise the cost would be too high for consumers.

CHAPTER 4 TEST - PROXIMATE ANALYSIS

1. What are the main components of all foods?
   A. Starch, roughage, protein, fat, and ash.
   B. Water, fat, fiber, carbohydrate, and ash.
   C. *Water, carbohydrates, protein, fat, and ash.*

All foods are made up of water, carbohydrates, protein, fat, and ash (what's left behind after a sample is completely burned). We'll study all of these in this course! Except for ash, which isn't horribly interesting.

2. How do you calculate how much water is in a food product from the side panel Nutrition Facts?
   A. *Add up everything else and subtract it from the serving size.*
   B. Weigh it before and after microwaving for 10 minutes.
   C. Subtract the total of the carbohydrates and protein from the total weight.

Water = serving size (g) - (fat + carbohydrate + ash + protein)

3. Nutrition Facts available on the side panel can help me:
   A. *Avoid eating too much fat.*
   B. Figure out if the food is going to taste good.

The Nutrition facts side panel has lots of information, if you care to look. In combination with the ingredients also listed on the side panel, you can gain a great understanding of what the food product contains, and make more informed decisions about what you want to buy or eat.

4. Ash is what is left once all organic material has been incinerated.
   True
   False

While we won't cover much about ash per se in this course, the components of ash are important in all the other systems they come from.

5. How do you calculate the amount of carbohydrates in a food?
   A. *Add up everything else and subtract it from the total weight.*
   B. Feed it to the author and see how much weight he gains.
   C. Determine the amount of sugar (carbohydrate) in the food.

While option 2 might work for me, please don't feed me any of my favorite carbohydrate loaded foods....I get enough on my own! Carbohydrate type and quantity can be determined experimentally using techniques such as gas chromatography (GC) and high-performance liquid chromatography (HPLC), but these are complex and expensive techniques. More simply the answer can be achieved by subtracting the sum of everything else, which are more easily experimentally determined, from the total weight of the food.

6. Proximate analysis means to:
   A. Water is of primary importance.
   B. use the side panel to analyze a food completely.
   C. "approximately" analyze all food components to find out what's in them
   D. <u>break a food down into it's five basic components.</u>

Sometimes also referred to as "composite" analysis.

7. A common element in protein used in laboratory tests to quantify how much protein is present in a food is:
   A. Carbon
   B. <u>**Nitrogen**</u>
   C. Oxygen

The Kjeldahl Method uses nitrogen as an indicator of how much protein is present.

8. Fat is soluble in:
   A. hot water
   B. salty water
   C. <u>some alcohols</u>
   D. ice cream

Fat quantities can be determined by dissolving fat into ethanol, evaporating the alcohol, and weighing the amount of fat left behind.

9. How many milligrams are in a gram?
   A. 10
   B. 100
   C. <u>*1000*</u>

"Mil" means "thousand"

10. Using the side panel on packaged foods can help me understand what's in my food and know if I am eating too much of some component that is not good for me.
    <u>True</u>
    False

The side panel is useful, especially for people who are supposed to avoid certain food components for health reasons.

CHAPTER 5 TEST - WATER

1. As oil solidifies slowly, it's density _____.
   A. stays the same.
   B. decreases.
   C. *increases.*

Almost everything's solid form is more dense than the liquid form, oil included. In other words, it shrinks as it solidifies.

2. As water slowly solidifies (freezes), the density _____.
   A. stays the same.
   B. *decreases.*
   C. increases.

Water is a strange and fascinating molecule because of hydrogen bonding. In the case of freezing, hydrogen bonds cause the crystalline structure to expand, rather than contract like most everything else in nature. So, ice floats!

3. Water activity is the partial pressure of water over a sample divided by the partial pressure of water over pure water. Choose the best option below you could use if you were trying to describe this principle to someone.
   A. Some food is really wet, and some is dry. You have to be careful when mixing them together.
   B. *Some foods hold onto their water more tightly than other foods. If two products are put together that don't have the same water activity, eventually water will move from one to another, sometimes with a poor result.*
   C. All things being equal, high water activity foods hold onto their water, and low water activity let go.

Just because a food appears wet doesn't mean it will give up it's water easily. You have to measure the water activity objectively (with an instrument) to find out where it falls, or look it up. Keeping water activity in mind will help you make sure your dessert is good today and tomorrow, rather than just today, for example.

4. What property of water makes it so good at cleaning?
   A. The water activity.
   B. *It's bipolar nature.*
   C. The equilibrium achieved when things are dirty.

Two hydrogens on one side of the oxygen result in a bipolar molecule, which makes it an excellent solvent for many things we must clean in everyday life.

5. Choose the best description of freezer burn.
   A. *Concentration of sugars, flavors, and oils, and loss of water at a food's surface, resulting in changes in texture.*
   B. When you put your tongue on a frozen popsicle, it gets stuck, and you pull it off, burning your tongue.
   C. Freezer burn makes food organoleptically yummy.

Freezer burn can destroy a food, but there are ways to prevent it. Much more on this in the full course, Introduction to Food Science: Water!

6. Flash freezing describes a very rapid freezing process. Compared to a slow freezing process, ice crystal size from flash freezing would be _____.
   A. larger.
   B. *smaller.*

The faster temperature is dropped below freezing, the smaller the ice crystal size will be. Flash freezing can sometimes prevent ice crystal damage to food products. In general, faster freezing is better on the quality of a food than slower freezing, for many reasons. More on this in the full course Introduction to Food Science: Water!

7. The Dew Point is the temperature at which air is saturated with water or, in other words, the air can hold no more water.
   A. *True*
   B. False

If the temperature goes any lower than the dew point, water will condense and fall out of the air as, for example, dew on the grass.

8. When making cooked food you intend to freeze, what are some things you need to consider. Mark all that apply.
   A. *Concentration of non-water components and unexpected flavor changes that might result (unexpected chemistry).*
   B. *Changes in texture.*
   C. Disruption of raw plant cells.
   D. *Freeze-thaw cycles common in today's freezers if it is going to be stored for a long time.*

Freezing foods is surprisingly complex. While it is a great way to preserve foods, food science can teach you how you can achieve a better rethawed or reheated food.

9. All other variables, such as pressure, as water changes phase, such as from a solid to a liquid, or a liquid to a gas, the temperature _____ until the phase change is complete.
   A. increases
   B. decreases
   C. *stays the same*

One of the great qualities of water is the constant temperatures it will hold as phases change. Then, the ability to change those temperatures by controlling other variables. An example is the popular Instant Pot, where the boiling temperature is increased by increasing the pressure. The constant temperature is also why ice chests work.

10. When an emulsion, such as mayonnaise, is frozen, usually the emulsion will _____.
    A. *break, meaning it will separate into its oil and water phases.*
    B. become homogenous, blending its oil and water phases.
    C. remain the same.

Without incorporating some Edible Knowledge, freezing emulsions usually results in poor results.

CHAPTER 6 TEST - CARBOHYDRATES

1. All sugars have the same sweetness.
   A. True
   B. *False*

Sugars vary widely in sweetness level. Food scientists can make great use of these when creating new products!

2. Choose the best answer that describes fructose, which is more hygroscopic than sucrose.
   A. *Fructose picks up moisture out of the atmosphere more readily than sucrose.*
   B. Sucrose picks up moisture out of the atmosphere more readily than fructose.
   C. You shouldn't use fructose in most food formulations, because it is hygroscopic.

Fructose attracts more moisture from the atmosphere, all things being otherwise equal, than does sucrose.

3. What's the primary difference between sugars and starches? Choose all that apply.
   A. *Sugars are sweeter than starches.*
   B. Sugars are longer chains of glucose than starches.
   C. *Starches are longer chains of glucose than sugars.*
   D. Sugars are carbohydrates and starches are not.

Starches have much longer glucose chain lengths than sucrose and are not considered sweet. One product made from corn starch, by shortening the chain lengths of starch, is glucose syrup. The shorter the starch is "cut", the sweeter than resulting product.

4. When bread crust browns and gives us those unique flavors we associate with baked bread, it is an example of:
   A. A good bread cook.
   B. *Maillard browning.*
   C. A call for butter and honey.

Maillard browning is responsible for many of the flavors we associate with baked bread and many other baked goods as well, including meats.

5. Maillard browning requires a _____ and a _____.
   A. *protein, reducing sugar*
   B. protein, non-reducing sugar
   C. starch, sugar

Protein and reducing sugars are necessary for maillard brown-

ing. In fact, that's the definition of it! These combine to create wonderful, robust flavors in slow baked or roasted goods, such as bread or roast meats.

6. Amylose is a _____ molecule than amylopectin.
   A. *Smaller*
   B. Larger

Amylose molecules are much smaller than amylopectin molecules.

7. Hydrocolloids are large molecules that absorb moisture, keeping a food moister, mimicking fat, and providing body and texture.
   A. *True*
   B. False

Hydrocolloids are amazing molecules, most from plants, that can help improve the texture, shelf life, and freeze/thaw stability of foods you make at home.

8. Examples of hydrocolloids are:
   A. xanthan gum, gum arabic, and cornstarch
   B. *xanthan gum, gum arabic, and guar gum*
   C. inulin, xanthan gum, and amylose

Most hydrocolloids are soluble fibers, meaning human digestive systems can't digest them. They all have their unique characteristics. The best thing is that many of them are commonly available for purchase by consumers. In encourage you to purchase some and try them out now that you have Edible Knowledge to back you up. It will take some experimenting, but will be worth it!

9. The primary difference between caramelization and Maillard Browning is:
   A. *How the reactions start since many of the end products (brown colors and flavors) are the same.*
   B. One requires protein, and the other requires sugar.
   C. They are not different.

Maillard browning and caramelization reactions can end up with similar compounds, but they get there in a very different path. The classic example is caramel candy. Starting with white cream and sugar, Maillard browning using protein from the cream and sugars from various sources (added and inherent to the cream) result in the brown colors and caramel flavor. To achieve colors and flavor through caramelization the water content must be reduced until there is very little left, even less than exists in caramel candies.

10. Cooked, hydrated potato starch gels are described as "long". This means that relative to most cornstarches there is _____ amylopectin.
    A. Less
    B. the same amount of
    C. *more*

A lower amylose/amylopectin ratio (meaning there is more amylopectin) means that the paste or gel will be "longer".

CHAPTER 7 TEST - PROTEIN

1. Protein levels of structure include:
   A. Simple, complex, and quaternary
   B. <u>*primary, secondary, tertiary, and quaternary.*</u>
   C. nitrogen, enzymes, and substrate.

   Primary (amino acid sequence), secondary (local three-dimensional structures), tertiary (complete protein three dimensional shapes), and quaternary (multiple protein interactions) levels of structure are all critical to the protein function.

2. The loss of structure in a protein due to heat, for example, is known as _____.
   A. desaturation.
   B. <u>*denaturation.*</u>
   C. consternation.

   Egg whites turning white when fried is an example of denaturation.

3. Glutenin + gliadin, with water and mechanical action results in a new protein complex called _____.
   A. <u>*gluten.*</u>
   B. glue.
   C. bread.

   Gluten is an amazing protein that makes many different kinds of foods possible, including bread.

4. Apples browning and onions producing chemicals that make our eyes water is an example of _____.
   A. nature in action.
   B. <u>*enzymatic action.*</u>

   Enzymes are critical to all life, but sometimes need to be controlled in a food product.

5. Are gluten-free foods better for everyone?
   A. Yes, because gluten causes problems for human intestinal tracts. That's why there are so many gluten free foods now.
   B. <u>**No. Only a select few particular diseases or sensitivities need to be concerned about foods that contain gluten.**</u>

   No. Since gluten comes primarily from wheat, everyone can eat it without any problems.

CHAPTER 8 TEST - FAT

1. If you wanted a high fat food to be less oily at room temperature, you would want the fat to be _____.
   A. liquid.
   B. *fully hydrogenated.*
   C. full of cis fats.

Hydrogenation converts a liquid fat to solid. Full hydrogenation ensures there aren't any trans fats, which are considered unhealthy.

2. Another way to source a fat that is more solid at room temperature is to choose one that has _____ fatty acid chains.
   A. *longer.*
   B. shorter.

Longer chain lengths generally result in more solid fats. There are sources of solid fats sourced from both animals and plants. Food scientists have lots of choices in order to make exactly what the customer wants.

3. All fats are built on a _____ molecule.
   A. Large
   B. Chained
   C. *Glycerol*

The glycerol molecule can have anywhere from zero to three fatty acid chains attached, which is what makes a mono, di, or tri-glyceride.

4. As fats age they can develop off-flavors and odors that most find offensive. When this happens the fat or food (containing fat) has become _____.
   A. *rancid.*
   B. extruded.
   C. nasty.

Rancid foods not only taste bad, they are also bad for you and can contribute to stomach cancer.

5. The following are ways that can reduce the onset of oxidative rancidity (choose all that apply):
   A. *Store in a cool place.*
   B. *Protect from light.*
   C. Flush with oxygen gas to help slow down the process.

Storing in a cool, dark place is a great idea. Flushing with oxy-

gen would accelerate oxidative rancidity.

6. Generally, donuts should be fried in a fat that is solid at room temperature.
   A. True
   B. False

Liquid frying oils would result in an organoleptically inferior donut.

7. The author believes we should avoid all trans fats.
   A. *True*
   B. False

The author chooses to eat foods in moderation, eating as close to nature as possible, as often as possible, but consuming more highly processed foods when the need or desired treat arises. He believes it is an easier and less stressful way to live, but encourages everyone to come up with their own food philosophy.

8. How many fatty acids does a triglyceride have attached to the glycerol molecule?
   A. One
   B. Two
   C. *Three*
   D. Four

Tri is the key. And, there are only three attachment points on a glycerol molecule, so four are not possible.

9. Potato Chips should be fried in fat that is solid at room temperature.
   A. True
   B. *False*

Chips fried in room temperature solid fat may have an unappealing appearance and mouthfeel. They should be fried in fat that is liquid at room temperature.

10. The mucus membrane irritating compound that results from fat thermal decomposition is called:
    A. acrimony
    B. *acrolein*
    C. smoke

Glycerol breaks down into acrolein, which irritates the eyes and nose, and also reduces finished food quality. Restaurants filter their oil to remove thermal fat degradation products.